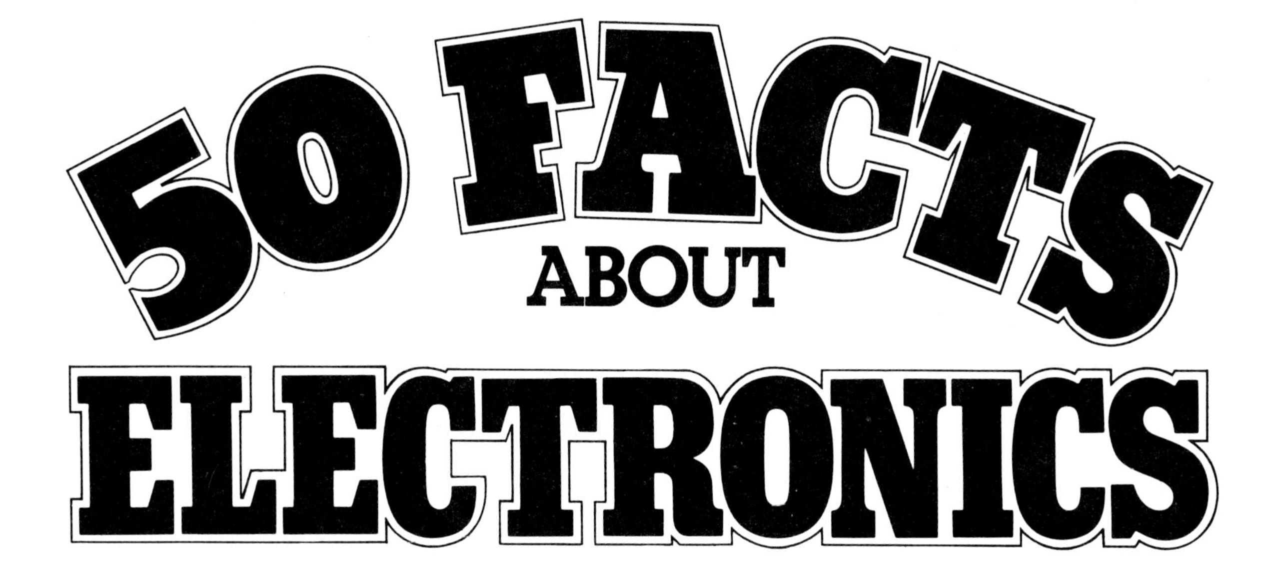

by
Mark Lambert

Piccolo
A Piper Book

CONTENTS

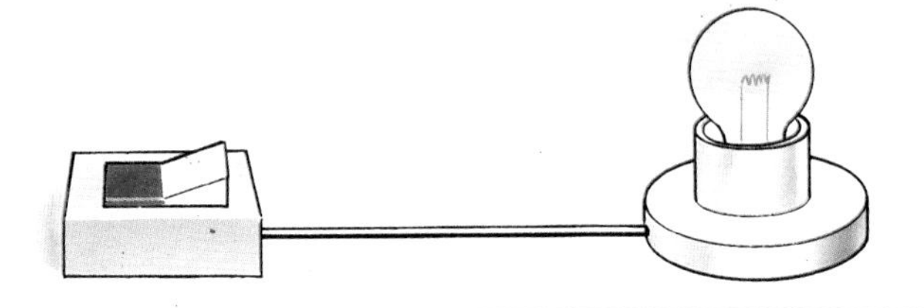

1 What is electronics?

In the modern world, electronic machines play an important part in our lives. And the number of different types of such machines is increasing at a tremendous rate. Just look around your home and see how many electronic things you can see. Radios, televisions, hi-fi equipment and pocket calculators are only a few of the things that are 'electronic'. You may also have some electronic games, or even a home computer – the ultimate electronic machine.

But what do we mean when we say that all these things are electronic? Electronics is simply the very accurate control of electric currents (see page 4). This control is achieved by using special devices, such as transistors (page 6), and resistors and capacitors (page 7). By using these devices, exactly the right amount of current can be directed to the right places at the right time.

The devices are often connected by wires. The path that the current follows as it travels along the wires from one device to another is called a circuit. The picture below shows some of the devices that you might find in a radio circuit.

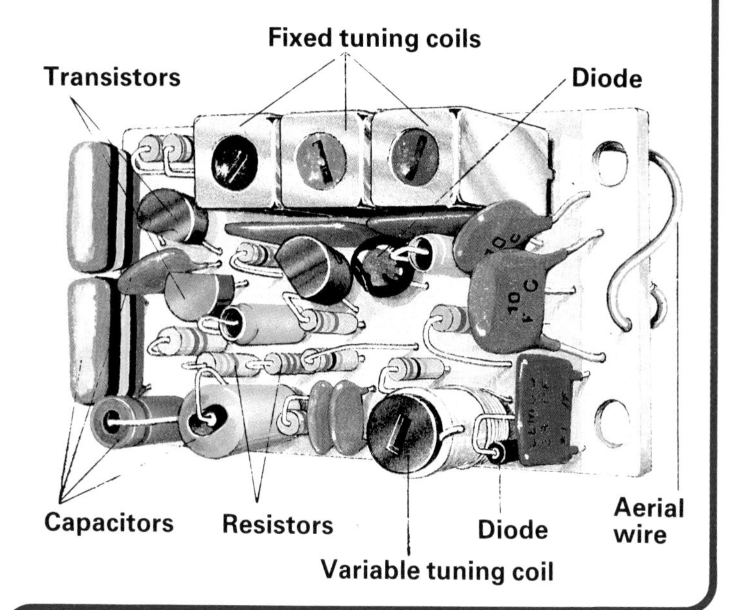

2 When did electronics begin?

Many people think that electronics began in 1883. In that year, the great American inventor Thomas Edison was trying to improve his new electric lamp. The glowing carbon filament (wire) in the lamp produced 'soot', which was blackening the inside of the glass bulb. Edison tried putting a metal plate inside the bulb, hoping to collect the 'soot' on it. He joined the plate by a wire to the electric circuit that sent current from a battery into the lamp.

To Edison's surprise, he found that a small current was passing inside the bulb from the hot carbon filament to the metal plate. Edison also found that this current only flowed when the plate inside the lamp was connected to the positive terminal of the battery.

Although Edison wasn't aware of it, he had made the first simple thermionic valve. This valve was at the heart of all early electronics, including the first computers, like ENIAC (shown here).

3 What are electrons?

Edison did not know how a small electric current managed to pass through his lamp. We now know that when the filament of the lamp was heated it became surrounded by a cloud of tiny negative electric particles called electrons. These electrons were attracted to the metal plate in the bulb when it was made positive. (Negative and positive charges attract each other.) So a current flowed through the empty space inside the bulb.

Everything in the universe is made up of tiny particles called atoms. But atoms themselves are made up of even smaller particles. At the centre of an atom is a nucleus – a bunch of particles called protons and neutrons. The nucleus is surrounded by

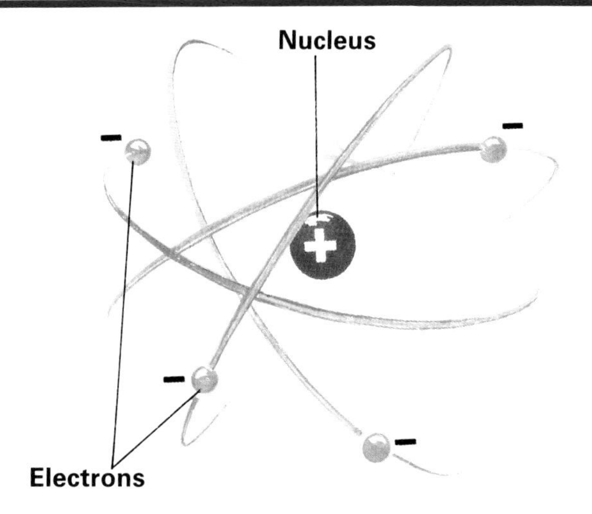

moving electrons. Neutrons have no electrical charge at all. But protons are positively charged and electrons are negatively charged. Under normal conditions each atom has equal numbers of protons and electrons, and the positive and negative charges cancel each other out.

What is an electric current?

An electric current can be thought of as the flow of electrons through a substance. The atoms of some substances have electrons known as 'free' electrons, that can be made to jump from one atom to the next. When a wire is connected from the positive to the negative terminals of a battery, the positive terminal attracts 'free' electrons from the atoms at the end of the wire to which it is attached. As each atom loses an electron, it becomes positively charged and therefore attracts an electron from its neighbour. This process is repeated continuously along the wire and so there is an overall flow of electrons from the negative end of the wire to the positive end. This is an electric current.

The nature of electricity began to be discovered during the 1700s. At that time it was thought of as an invisible fluid. The word 'current' describes the movement of water, so the movement of electricity

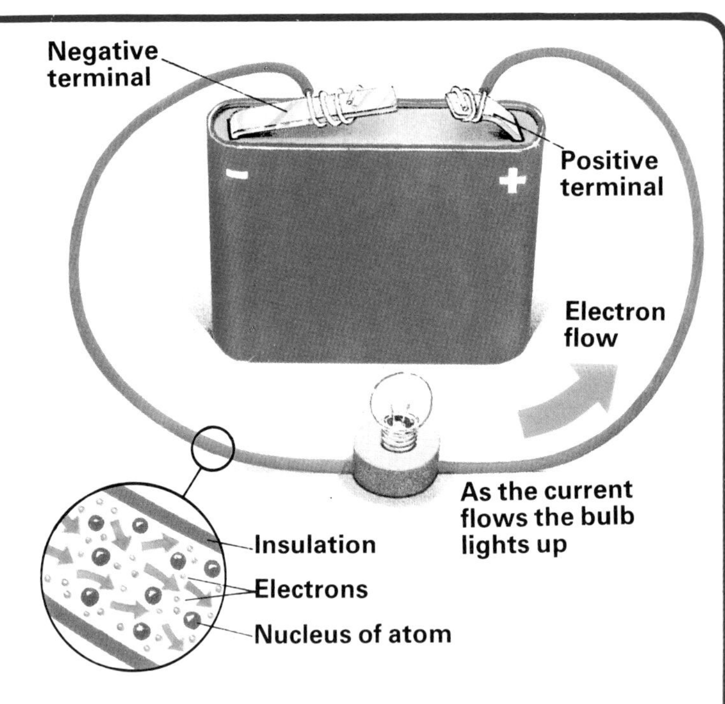

became known as 'electric current'.

Some materials allow electricity to pass through them more easily than others. In general, good conductors of electricity are those with plenty of 'free' electrons, such as metals. Materials such as rubber and plastics have no 'free' electrons and so are non-conductors, or insulators.

What are valves?

A simple electronic valve contains two electrodes. These are metal conductors that are only attached to a circuit at one end, and either give out or receive 'free' electrons. A thermionic valve works by passing a stream of electrons from a heated cathode (a negative electrode) to an anode (a positive electrode).

There are two main kinds of thermionic valve. A diode valve has two electrodes – an anode and a cathode. Electrons flow only from the cathode to the anode and so a diode passes current in one direction only. If alternating current (current that changes direction many times every second) is supplied to a diode, it will be turned into pulses of direct current (current that always flows in the same direction). This process is known as rectification.

A triode valve is like a diode, but it has a third electrode known as the grid. A triode can be used to switch the flow of current off or on in a circuit, or to amplify (strengthen) it.

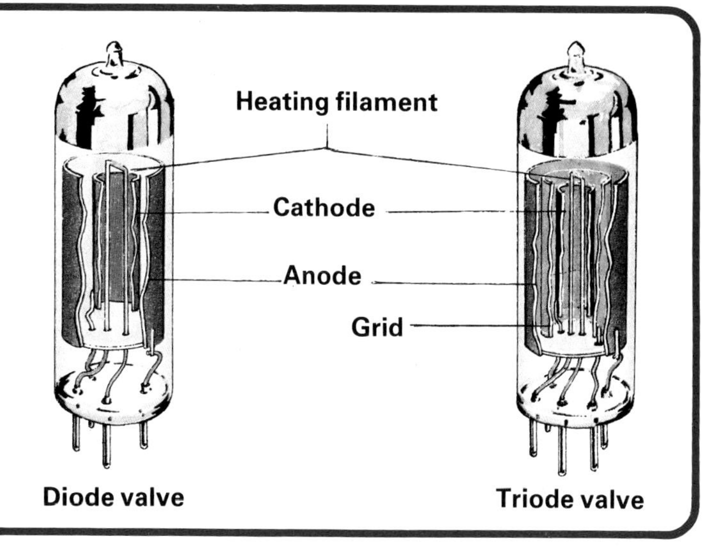

How were transistors invented?

Thermionic valves are large, easily damaged and eat up a lot of electricity. They also take some time to warm up. During the 1940s a team of American scientists, William Shockley, John Bardeen and Walter Brattain of Bell Laboratories, were trying to find an alternative device, using the new semiconductor materials that had been discovered. In 1947 they succeeded in making the first transistor. Over the next ten years they perfected this device and since then transistors have become smaller and smaller.

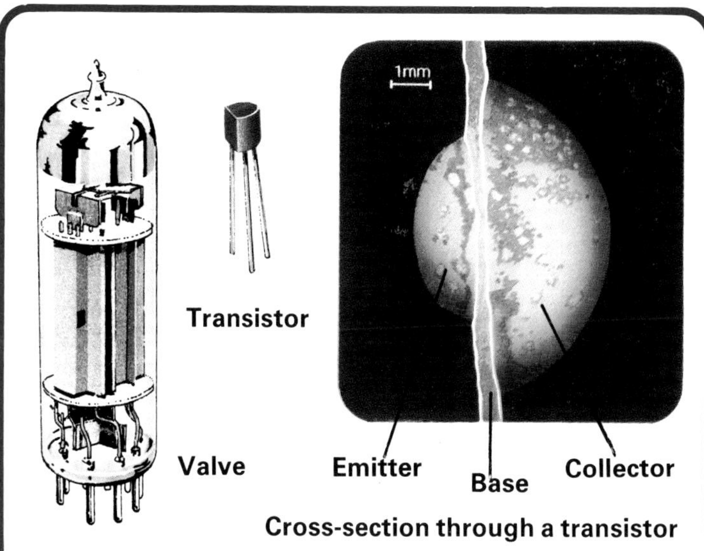

Cross-section through a transistor

Semiconductors are materials that can be both conductors and insulators. At low temperatures they are insulators, but at higher temperatures some of their electrons gain enough energy to become 'free'. Two materials that behave in this way are germanium and silicon.

Pure semiconductors are not generally very useful because they need heating. However, a semiconductor can be made to conduct electricity much more easily by adding a small amount of another material to it. This is known as adding an impurity. Modern semiconductors are usually made of silicon mixed with small amounts of impurities such as phosphorus or aluminium.

Semiconductors are now used in diodes as well as in transistors, and these two devices have now largely replaced the old valves.

7 How does a diode work?

We call semiconductor materials used in diodes n-type and p-type. An n-type material contains impurity atoms (red in the drawing) that each provide an extra electron. A p-type material contains impurity atoms (in green) that each have one less 'free' electron than the surrounding silicon atoms (in black). The result is that positive 'holes' (empty spaces) are left between some of the atoms.

A diode passes current in only one direction. If the p-type end is connected to the negative terminal of a battery and the n-type end is connected to the positive terminal, the electrons and 'holes' both move away from the centre. This creates a region across which electrons cannot flow. However, if the battery connections are reversed the electrons and 'holes' move towards each other and the electrons (current) flow through the diode.

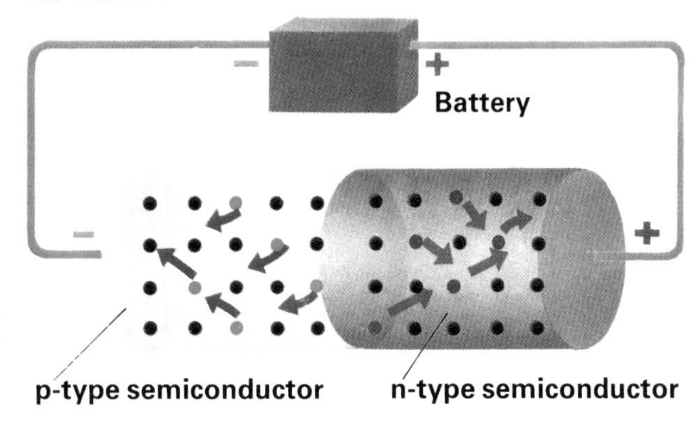

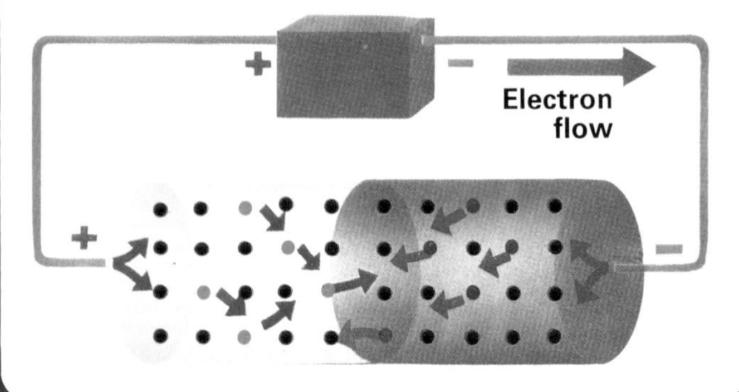

8 How does a transistor work?

A transistor is made of a thin slice of semiconductor sandwiched between two thicker slices. An n-p-n transistor has a layer of p-type material between two layers of n-type material. The layers are known as the collector, base and emitter. If a battery is connected to the transistor as shown in the top drawing, electrons flow towards the positive terminal. However, as in a diode, a barrier soon builds up at the collector-base junction and current cannot flow.

To allow current to flow, it is necessary to connect a low power source between the base and the emitter to push a few electrons into the base. This then allows electrons to flow from the emitter to the collector.

A transistor can be used as a switch – if the base current is switched off the main current stops. And as an amplifier – a small varying current at the base gives a larger varying emitter-collector current.

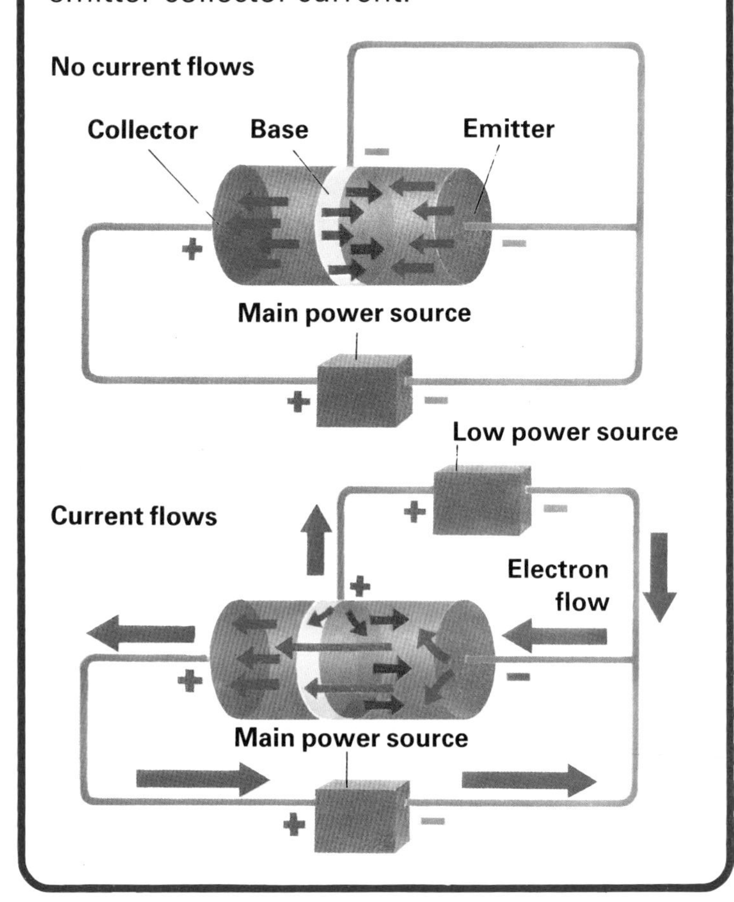

What is a resistor?

A resistor, as its name implies, resists the flow of current – it slows down electrons. All materials resist the flow of current to some extent and the amount by which a piece of material does this is called its resistance, and it is measured in ohms (Ω). A piece of copper has a very low resistance, and is a good conductor. A piece of plastic, however, has an extremely high resistance.

Resistors are used in electronics to control the amount of current flowing in various parts of the circuits. A resistor usually consists of a core of carbon, which is a poor conductor, inside a layer of insulator. Its resistance is shown by three colour-coded stripes at one end. The colour codes are shown in the chart. The first stripe indicates the first number and the second stripe indicates the second number. The third stripe indicates the number of 0s (zeros) that follow these two numbers. The stripes on the first of the resistors shown here are brown (1), black (0) and red (2 zeros), which indicates that the resistance is 1000 ohms. These are fixed resistors. There are also variable resistors, whose resistance can be altered by turning a knob.

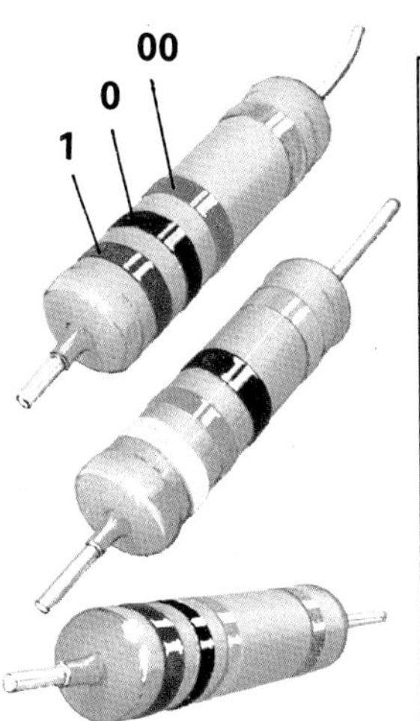

	Colour	Number or Number of 0s
	Black	0
	Brown	1
	Red	2
	Orange	3
	Yellow	4
	Green	5
	Blue	6
	Violet	7
	Grey	8
	White	9

What are capacitors used for?

A capacitor has two metal surfaces, or plates, separated by an insulator. When a battery is connected to it, one surface becomes positively charged and the other becomes negatively charged.

A fixed capacitor can be used to smooth out the pulses of direct current given out by a diode that is receiving alternating current (see page 5). The capacitor stores the bursts of current from the diode and lets it out as a more even flow. But a capacitor can also work with alternating current. In a radio, hundreds of radio waves of different frequencies (see page 8) pass from the aerial to a tuning circuit, where they produce tiny alternating currents. By adjusting the position of the metal plates on a variable capacitor, the tuning circuit can be made to receive one particular wave frequency.

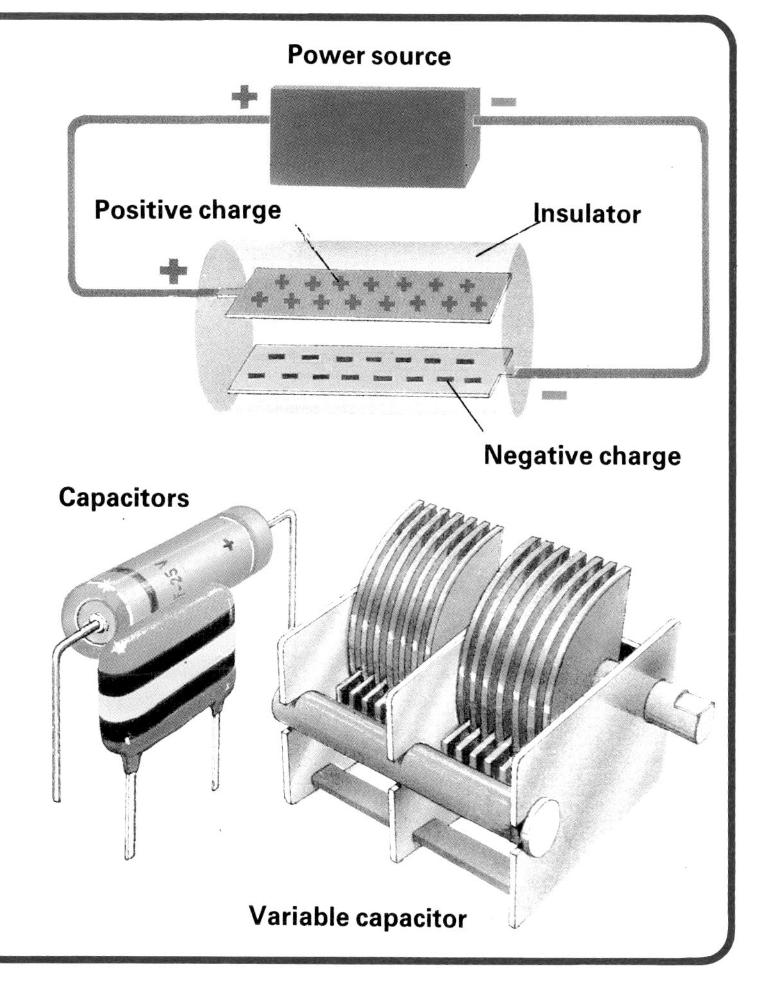

What are radio waves?

Radio waves are one form of electromagnetic energy. The other forms are shown below. Like the waves in the ocean, all these forms of energy have a wavelength (the distance from one crest of a wave to the next) and a frequency (the number of waves that pass a given point in one second). Radio waves vary from UHF – ultra high frequency and short wavelengths, to 'Long wave' – low frequency and long wavelengths.

A powerful alternating electric current can be made to produce radio waves. Sounds and TV pictures can also be turned into varying electrical current. This current can then be used to alter or modulate the radio or carrier waves, which carry the electrical 'pattern' of the original sound or picture. Sound is often transmitted by frequency modulation (FM), where the frequency of the carrier waves is made to vary slightly. TV pictures are transmitted by amplitude modulation (AM), where the height of the carrier waves is made to vary.

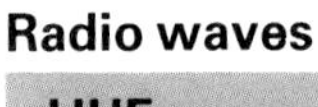

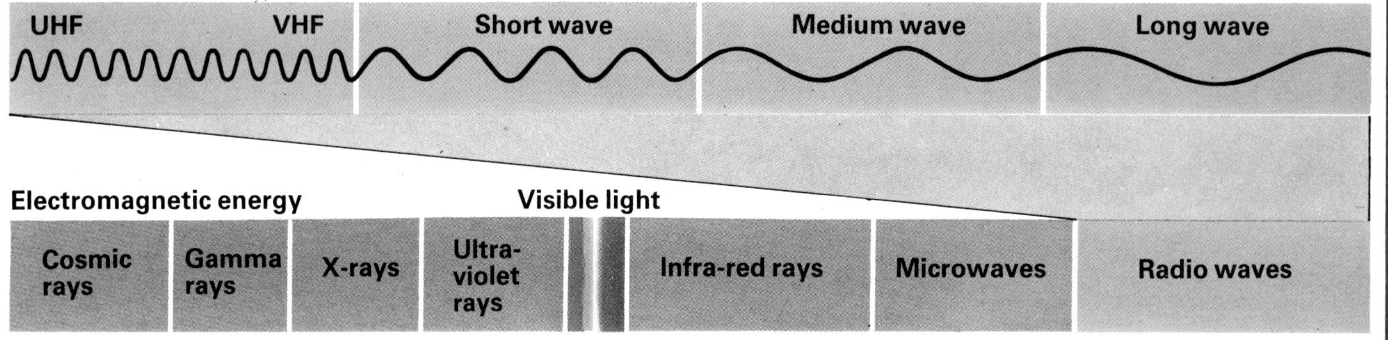

12 How does a radio work?

The simplest kind of radio picks up radio waves, tunes into those with the required frequency and passes them on to an earphone. Radio signals are received first by an aerial. They are passed on to a part of the circuit that contains a wire coil and a variable capacitor. This selects the desired frequency from the hundreds of signals picked up by the aerial.

Next comes the diode. This rectifies the selected signal; in effect it removes one half of the carrier wave, leaving a series of varying pulses of direct current.

Finally, a capacitor in the circuit acts as a filter, or demodulator. It removes the carrier wave, leaving only the sound signal that it carried. This is passed on to the earphone, from which the sounds are produced.

More complicated radios may have several tuning circuits, each covering a particular band of frequencies. They also have amplifying circuits, which contain transistors, to strengthen the signal.

Aerial wire
Coil
Diode
Capacitor
Ferrite rod
Variable capacitor
Socket
Jackplug
Earth wire (to kitchen tap or iron rod in the ground)
Earphone

13 How does colour television work?

Television works by changing light waves into electrical signals. This is done inside the television camera. When a camera is pointed at a particular scene, the light waves from the scene are picked up by the camera lens and passed through a series of coloured mirrors. The mirrors separate the light waves into three colours – green, red and blue. The light waves for each colour then pass through another lens to a separate camera tube. Inside each tube, a rapidly moving beam of electrons turns the light waves into electrical signals.

The signals from the three tubes are then added together to make one signal, which also carries a code for each colour. The signal then goes to the transmitter.

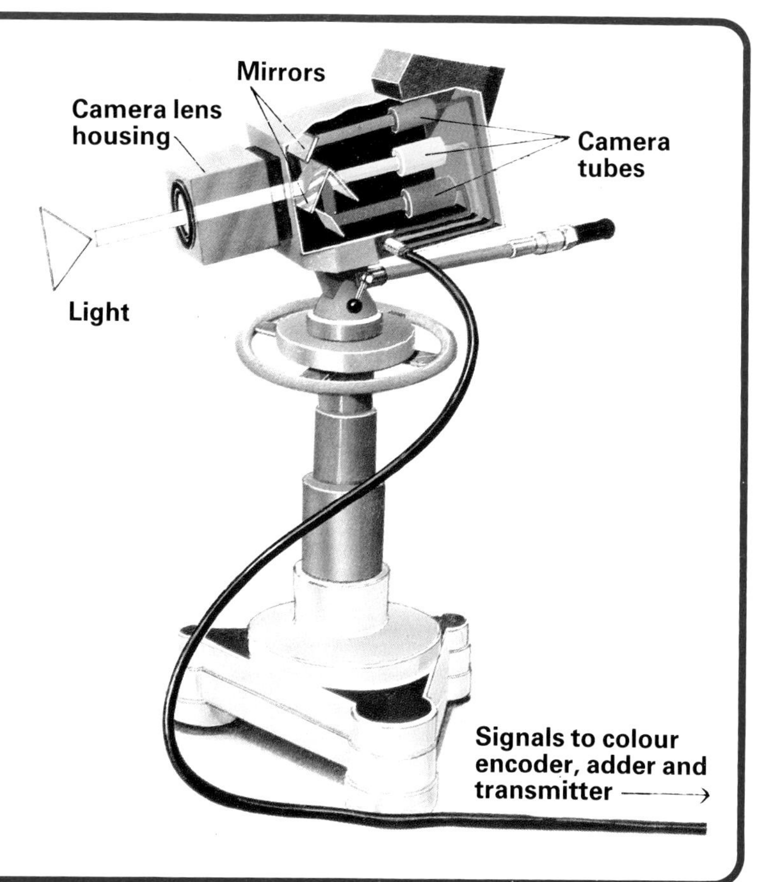

14 How do pictures travel through air?

Television pictures and the sound that goes with them are carried by radio waves – usually VHF (very high frequency) or UHF (ultra high frequency) waves. From the cameras and microphones in the studio the sound and picture signals are amplified and go to the transmitters. There they are broadcast on carrier waves.

High frequency waves cannot travel through mountains or buildings, so television signals are often picked up, amplified again and retransmitted by relay stations. When they reach your television aerial they pass to the set, which contains devices for separating the sound signals from the picture signals and amplifying them. Finally, the signals are demodulated (the carrier waves are removed) and the original sound and picture signals pass to the loudspeaker and your screen.

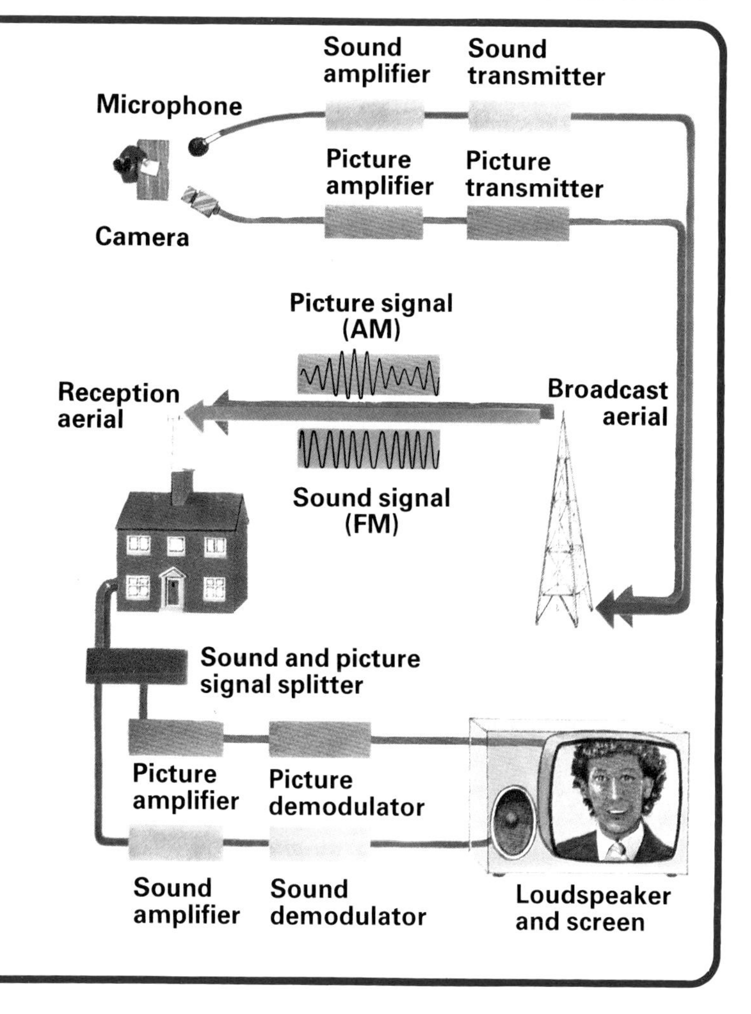

15 What does a television set do?

Inside a television set there are three electron guns, one for each colour. The picture signal is split into three beams of electrons that scan (move to and fro) across the screen at very high speed. Between the electron guns and the screen is a plate called the shadow mask, which has thousands of tiny holes. As the electron beams move across each hole, they pass through and strike the screen. This is covered with tiny red, green and blue phosphor dots, which glow when struck by the electrons.

The strength of each electron beam is controlled by the picture signal. A weak part of the beam (containing few electrons) results in a darker colour dot. So the picture on the screen is made up from light and dark coloured dots.

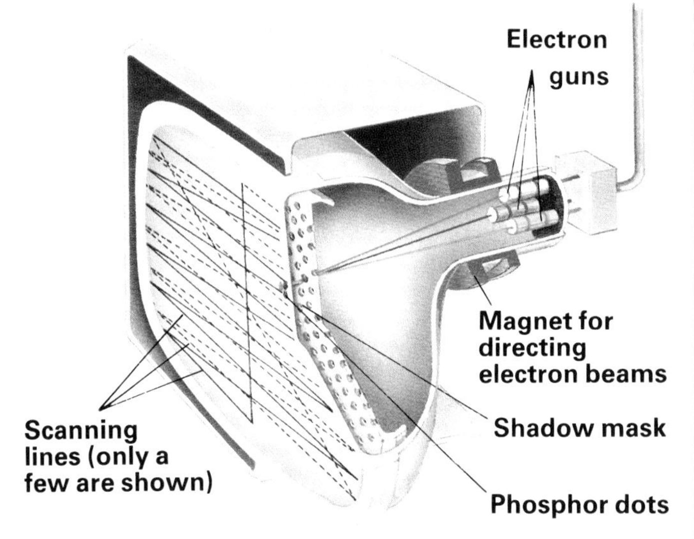

16 What is a video player?

As well as being broadcast immediately, picture signals can also be stored on film or magnetic tape, or even on a disc (like a record), and then played later. Storing pictures, as well as sound, on tape or disc is known as video recording, and the machines that are used to play them back through a television set are called video players. A video player which uses tapes can record programmes from your television as well as play them. But video discs (again like records) can only be played; you cannot record things on them at home.

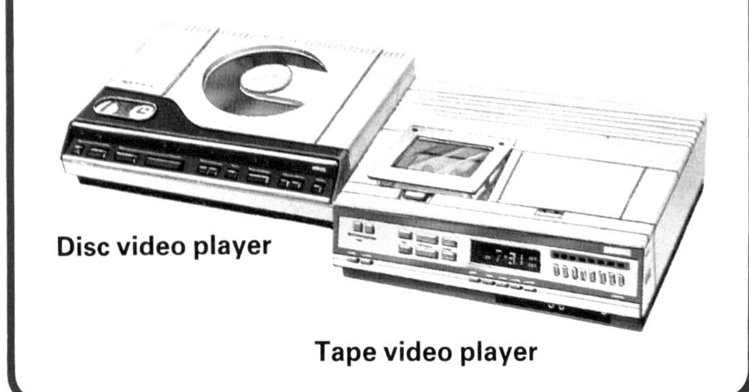
Disc video player

Tape video player

17 What is a hi-fi system?

No system that uses tapes or discs to play recorded music has yet managed to produce a 'perfect' imitation of the original sounds. If a recording is fairly close to the original it is said to have 'high fidelity', and the system which plays it is known as a hi-fi. A typical hi-fi has a turntable (for records), a cassette player (for tapes), a radio tuner (to pick up radio), an amplifier and loudspeakers. Hi-fi systems are constantly being improved with new electronic devices. The latest of these uses a laser to 'read' the music from a special 'compact disc'.

18 How do we get live news?

At one time all television news items were recorded on film. Then the film had to be sent back to the studio to be processed, edited, projected and finally broadcast – all of which took a lot of time. With the development of portable video equipment, the newspeople now have a faster method of collecting and transmitting news. It is known as electronic news gathering (ENG).

News events are recorded by hand-held video cameras onto tape carried in lightweight, portable video recorders. Sometimes, like film, the tape is sent back to the studio. But ENG vans carry a portable aerial, and pictures from the video camera can be broadcast 'live' – straight from the ENG van to the television station and onto our television screens.

19 What is an electronic bug?

Intelligence agents and industrial spies gather large amounts of information by listening in to other people's conversations. As electronic parts get smaller and smaller, it has become possible to produce highly efficient tiny electronic radio transmitters, or bugs. These pick up sounds and send them to a receiver some distance away.

Most electronic bugs are made for a particular job. A telephone bug needs no battery or microphone (it uses the power and microphone of the telephone itself) and the smallest type is no larger than a grain of corn. But even those listening devices that do need batteries and microphones can be as small as a lump of sugar. And they are often cleverly disguised or concealed. Bugs carried on the body can be made to look like pens, watches, tie-pins, lapel badges and cuff-links. Tiny pipe-shaped bugs can be put through keyholes, and hidden in walls and floors. One clever type of bug looks and works exactly like an ordinary light bulb.

Bugs can usually be discovered by devices that react to the radio frequencies the bugs produce. But new technology will soon make it possible to produce bugs whose signals are much more difficult to detect.

20 What is radar?

Radar is used to discover the presence and position of distant objects. Ships use radar to find other ships at sea. Airports use radar to check the positions of aircraft.

The word 'radar' means **ra**dio **d**etection **a**nd **r**anging. A transmitter/receiver sends out bursts of very short wavelength radio waves from a continually turning aerial (called an antenna). If the radio waves hit an object, they bounce back to the aerial. Using the time difference between sending and receiving the signal, and the position of the aerial, the receiver sends out an electrical signal which contains information as to the direction of the object and how far away it is.

The signal is sent to a tube (like a television tube), and the object appears as a glowing 'blip' on a radar screen. From the position of the blip on the sceen, the radar operator can see what the distance and direction of the object is.

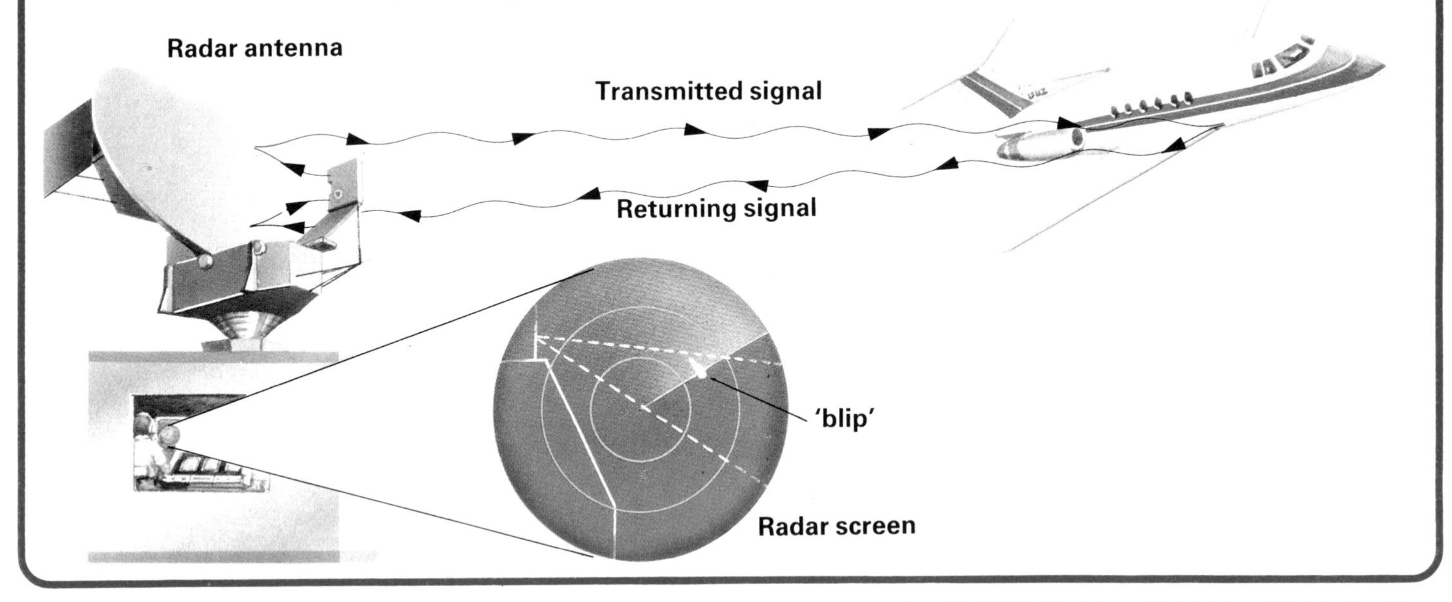

21 What are synthesizers?

A synthesizer is a very complicated electronic musical instrument. Most musical instruments produce sound as a result of something, such as a string or a column of air in a tube, being made to vibrate. But an electronic instrument has no strings or tubes or air. Instead, electronic circuits are used to generate electrical signals which are then fed to an amplifier and loudspeaker. The keyboard and other controls on a synthesizer act as switches to select the required electronic circuits.

Synthesizers can be made to produce sounds that no other instrument can make.

22 What is a printed circuit?

A printed circuit is simply a network of copper strips attached to a board. They can be used in all kinds of electronic equipment.

To make a printed circuit, a plan of the circuit is copied, through carbon paper, onto a clean copper plate glued to an insulating board (1). The carbon lines on the copper are then painted over with a substance that will resist being etched (eaten away) by acid. The board is put into a bath of acid or etching solution and all the copper, except for the painted lines, is etched away (2). The printed circuit is removed and cleaned, and holes are drilled at various points along the copper lines (3). Finally, the electronic devices, or components, are placed on the non-copper side of the board. The wires from each component are fed through the holes and soldered (joined up) to the copper (4). Current can now flow along the lines to each component.

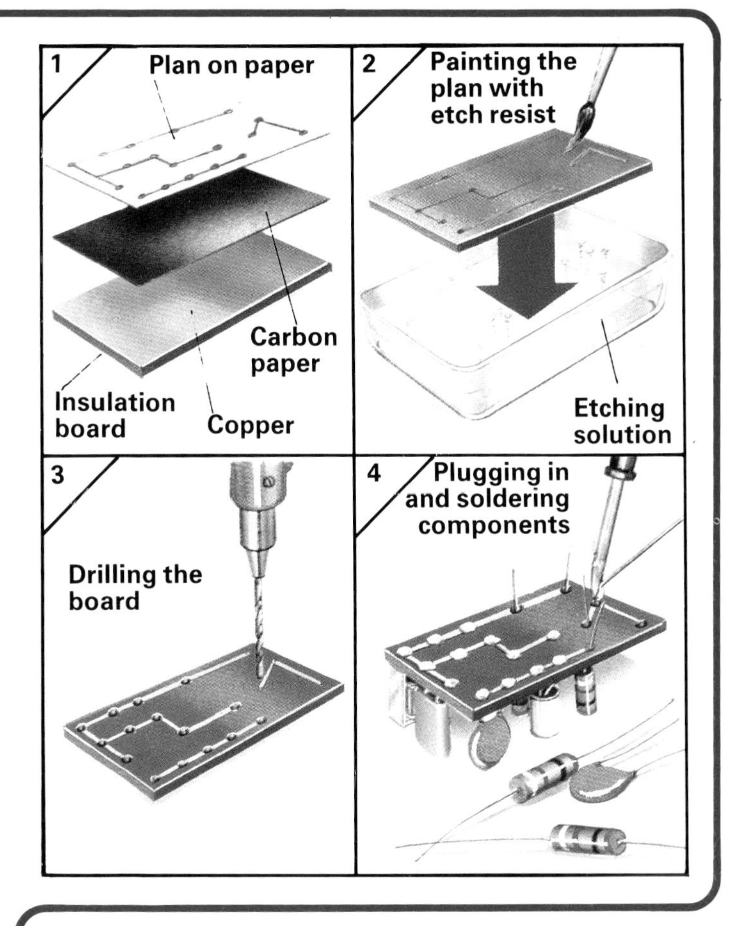

23 What is a silicon chip?

A silicon chip is a tiny slice of silicon material (about 5 mm square) which contains hundreds or even thousands of electronic components. Because silicon is a good semiconductor (see page 5) it is used to make transistors and diodes. But in the late 1950s, it was discovered that it was possible to form circuits of microscopic transistors and other components directly onto the surface of a silicon chip.

Because silicon chips are so tiny, it is possible to build extremely complicated machines that are still small in size. The microcomputer, for example, could not exist before the invention of silicon chips. A single chip in a plastic case can control all of the major functions of a small computer. And when a number of chips are plugged into a printed circuit board, they do the same job as a roomful of ordinary components.

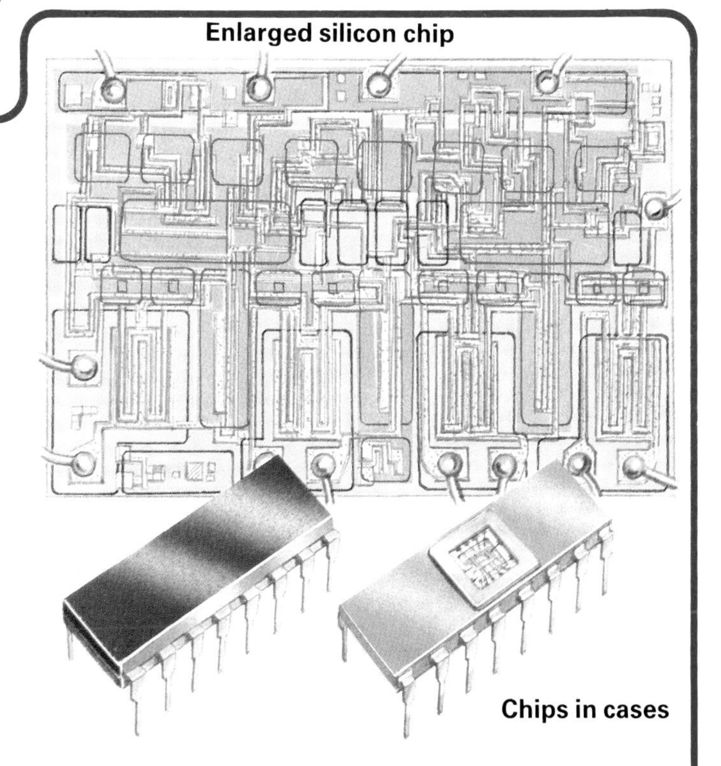

24 How are silicon chips made?

A large crystal of pure silicon is formed in a furnace (1). This is sliced into thin wafers, which are polished smooth (2). Meanwhile, the circuit pattern is designed. The pattern is separated into layers. Each layer is reduced to the size of a chip and printed onto a photographic mask. Each mask can carry hundreds of prints of the same layer of circuit (3). The masks are printed onto a wafer. The pattern of each layer is etched into the wafer and impurities are added inside a furnace (4).

The finished circuits on each wafer are tested (5). Many of them will be faulty. The final wafer (6) is cut up into chips. The good chips are wired up with very fine wires to metal contacts which carry current to and from the chips (7). Finally, the chips are mounted in plastic cases (8).

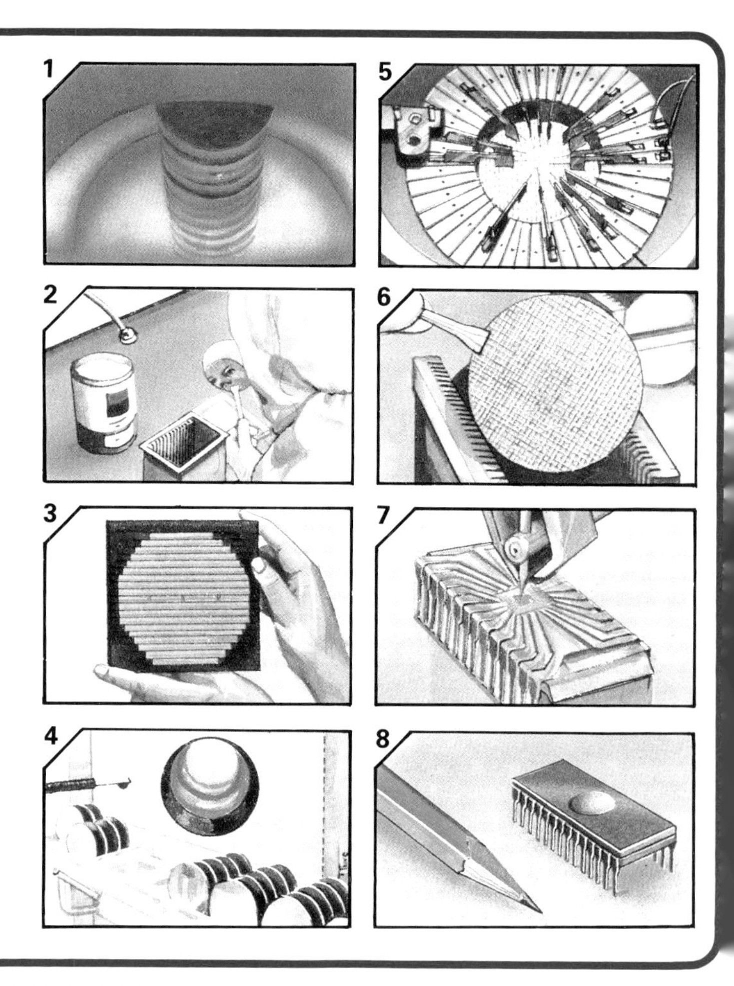

25 How does a calculator work?

We use calculators to do our sums for us, and some of them even have simple games built into them or can play music as well.

A pocket calculator can do all these things because it contains at least one silicon chip (or microchip). When the calculator keys are pressed, contacts are made and electrical signals are sent to the chip. Here, the signals are changed into a special code called binary code (see page 19), which is made up of groups of electrical pulses. The chip already has instructions about how to add, subtract, multiply or divide binary codes built into some of its circuits. So it can work out your answer in binary code and then send the answer back, as electrical signals, to the display unit. This shows the signals as numbers on a screen.

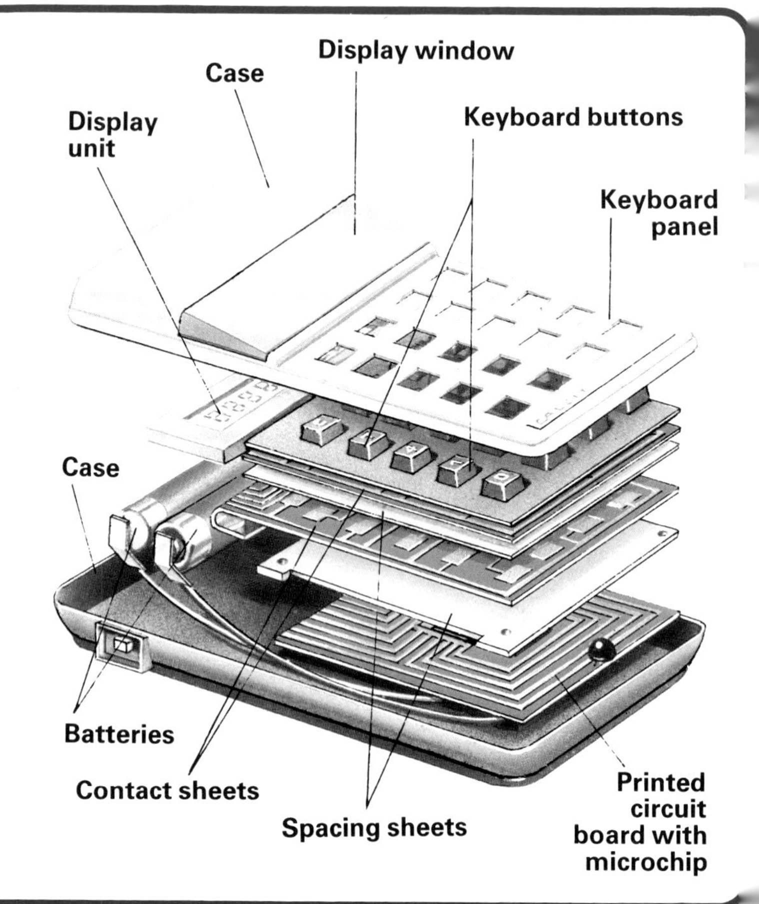

26 How do electronic displays work?

If you look closely at the numbers displayed on the screen of a calculator or a digital watch, you can see that each number is made up of separate segments. The maximum number of segments needed to make up a number is seven. Each number uses a different combination of some or all of these segments.

In the earliest calculators, each segment was a light-emitting diode (LED). LEDs glow like tiny bulbs when current passes through them. However, LEDs use a lot of battery power and most calculators now use liquid crystal displays (see below). Most liquid crystal displays still use the seven-segment system, but these are slowly being replaced by liquid crystal dot displays. These contain several rows of tiny dots. By activating the right combination of dots it is possible to show letters as well as numbers.

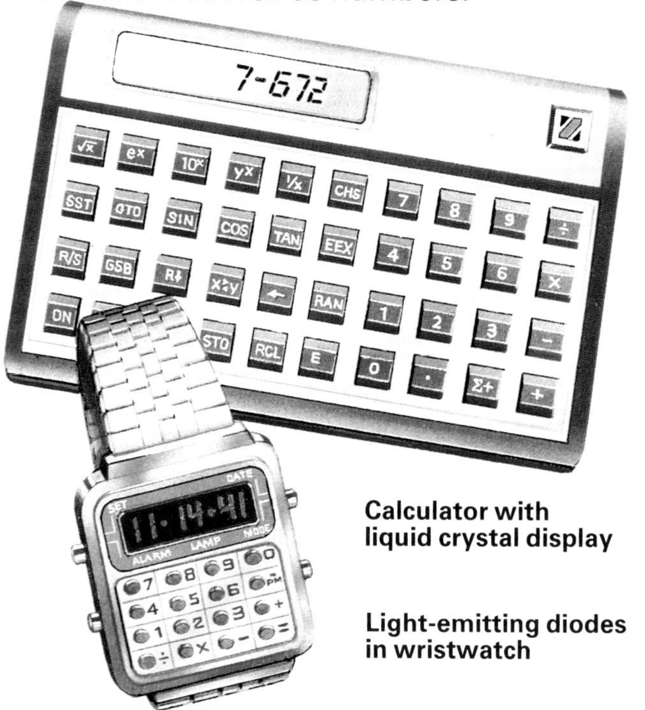

Calculator with liquid crystal display

Light-emitting diodes in wristwatch

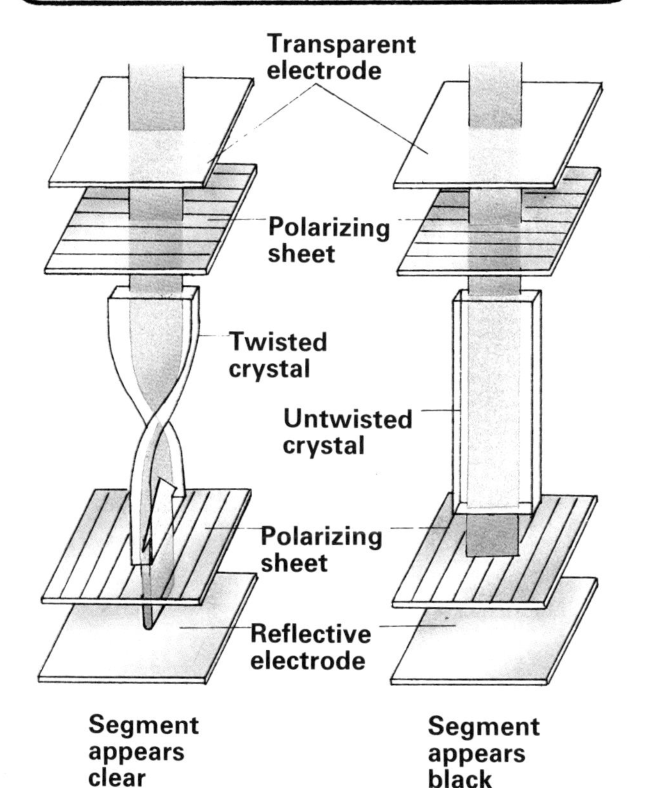

27 What are liquid crystals?

Like all crystals, liquid crystals are made up of molecules (groups of atoms). But the arrangement of the molecules can be altered by electricity; changing the crystal's shape.

A liquid crystal display contains a number of these crystals (one for each segment). When there is no current passing through them, the crystals are twisted. Light from the display window passes through a transparent electrode and a polarizing sheet, which directs the light to the crystals. The light is twisted by the crystals so that it can pass through a second polarizing sheet to a reflecting electrode. This sends the light back to the display window and the segments appear clear. But when current is passed through a crystal it untwists. The light cannot now pass through the lower polarizing sheet. It is not reflected back, and so the segment appears black.

28 What is a computer?

A computer is an electronic machine. But this type of machine has changed our lives during the last fifteen years. A computer can store, compare and process information by performing calculations at lightning speed – many thousands of times faster than the human brain can.

The first fully electronic general-purpose computer was built in 1946. Known as ENIAC, it filled a large room and used row upon row of thermionic valves (see page 3). These ate up a large amount of power and tended to overheat and burn out.

The problem was solved by the invention of the transistor. It soon became possible to build much smaller, more reliable computers. Today's computers use little power and come in many sizes. The largest are the supercomputers, which are used for such tasks as weather forecasting. Mainframe computers are used by large organizations for such things as calculating wages, sending out bills and controlling complex industrial processes. Mini-computers are smaller versions used by businesses, hospitals and small factories. Microcomputers are often used as personal computers by office staff. They are also used in schools and in the home.

29 Why do we need computers?

Many people find computers difficult to understand. Others see them as a threat to human jobs and skills. But computers are important in the modern world because of the way in which they can handle vast amounts of information in a very short time. Much boring work has been given to computers, leaving people free to do more important things. Computers have given us more efficient banks, shops, offices and factories. The picture here, for example, shows a computerized supermarket till with built-in stock control – a great improvement on notepaper and pencil. And some things, such as space research and exploration, would be impossible without the help of computers.

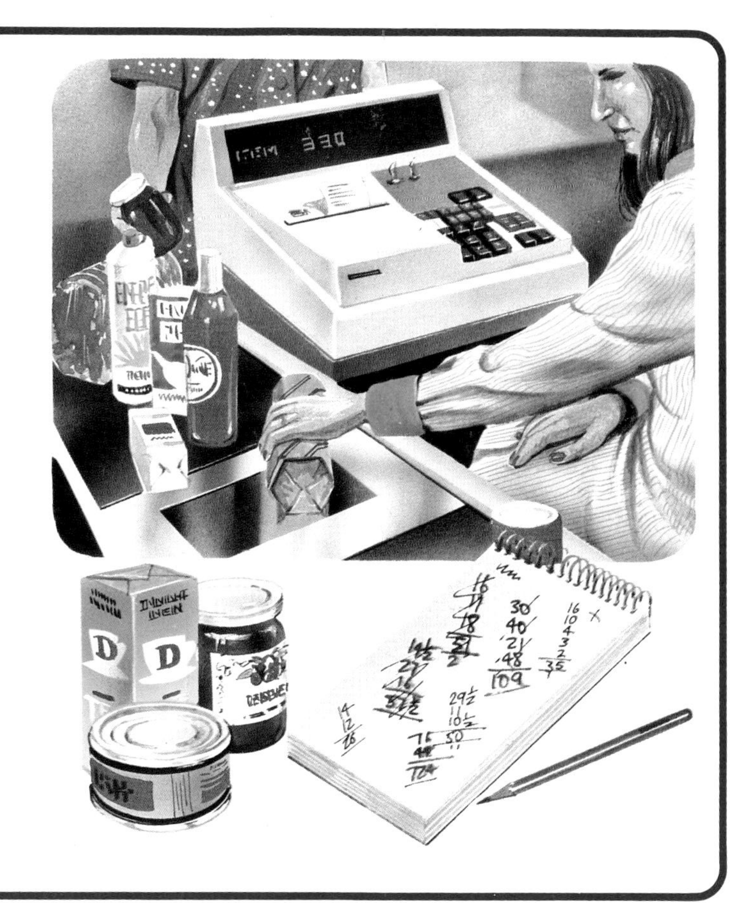

30 How does a computer work?

A computer consists of several distinct units. We put information into the computer via an input unit, which is often a keyboard. This is linked to a central processing unit (CPU), which carries out the task of working on the information (see page 18). The CPU is controlled by the main memory (see page 20), which contains instructions for handling the information. The main memory also stores the program, a special list of instructions which are fed in via the input unit, telling the computer exactly what we want it to do. Additional information that the computer may need to work on may be held in an extra memory, or backing store. Finally, the result of the computer's calculations are fed to an output unit, such as a television screen or printer (see page 18).

Special-purpose computers such as those found in washing machines or computerized sewing machines, have programs stored permanently in their memories. Such computers can only do certain things. In general-purpose computers, on the other hand, the programs and other information are kept separate. This is known as software. The electronic and electrical equipment – the input and output units, the CPU, and the memory stores – is known as hardware.

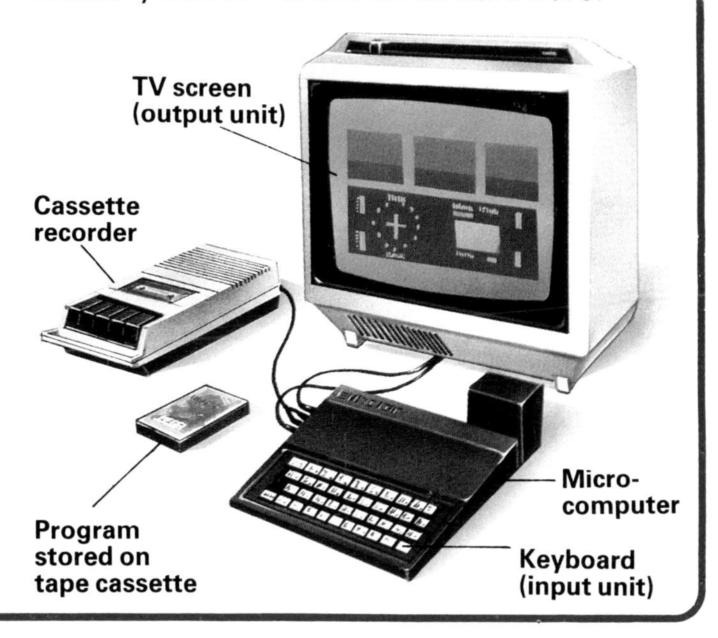

31 How are computers used?

Information can be fed into and out of a computer in a number of ways, and the machines which do this are called input and output units. Words and numbers can be typed in, using a keyboard. This information can be seen on a screen and, by connecting the computer to a printer, it can be printed out onto paper. A light pen is used to draw pictures directly onto the screen, and games paddles are used to move images around on the screen. Less obvious devices are things like the sensors and motors in a robot arm. The sensors feed information to the computer about the position of the arm, and the motors move the arm to the next position.

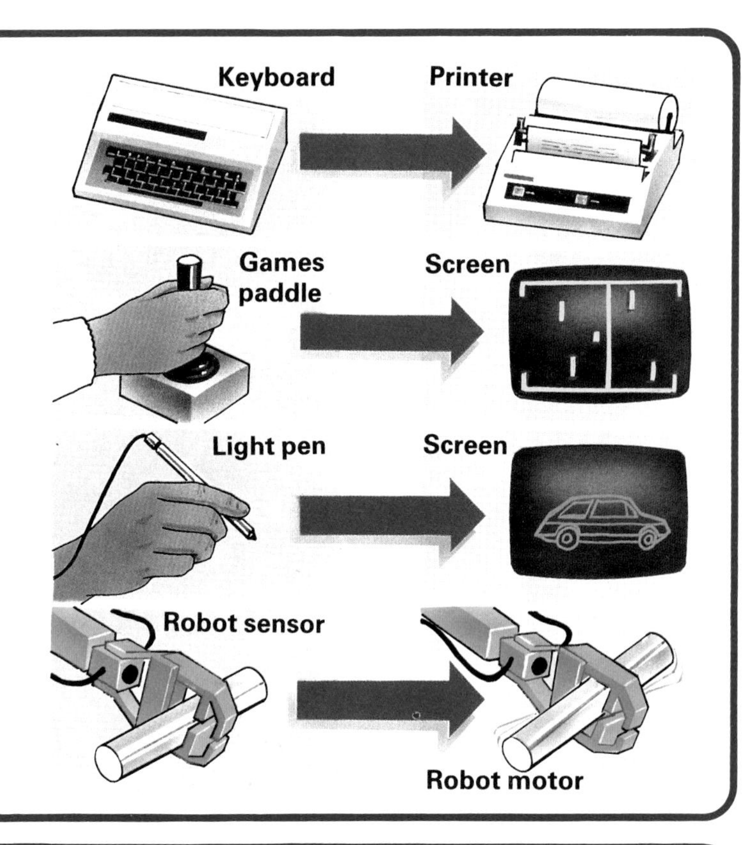

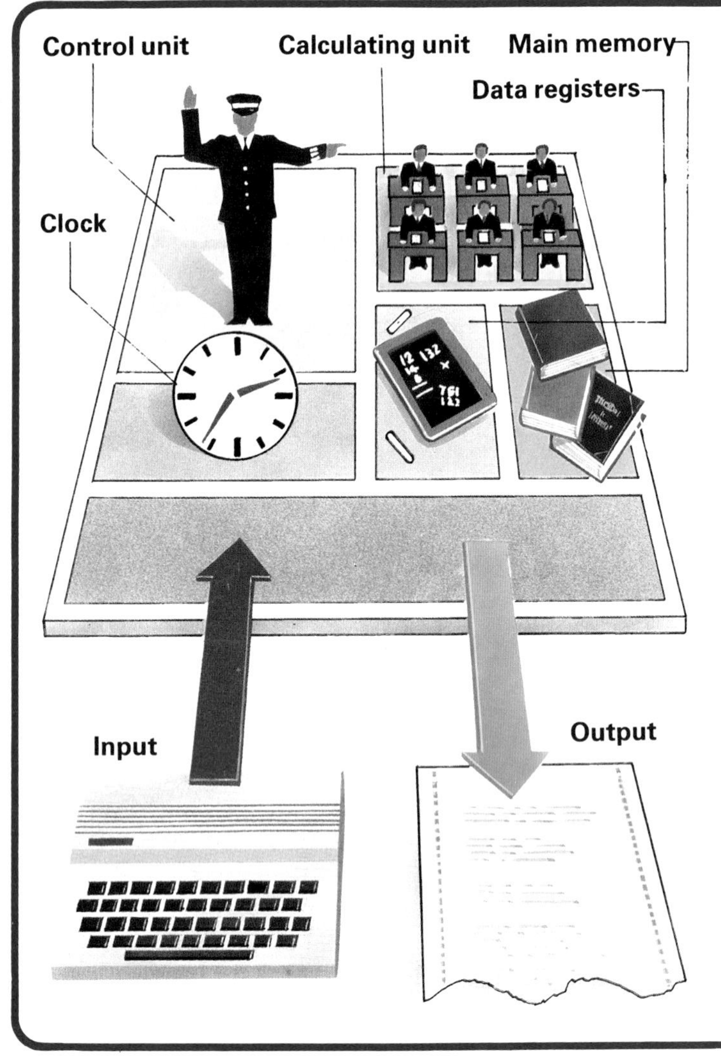

32 What is the heart of a computer?

All calculations and information processing are carried out by the central processing unit (CPU), helped by the main memory. The various parts of the CPU have to carry out their tasks at exactly the right time. To achieve this, the control unit uses an electronic clock to regulate the pulses of current that pass around the CPU.

The program and other instructions are held in the main memory. One by one, each stage of the program is taken from its place (address) in the memory. The control unit then finds the information (data) that is to be operated on; this may be in the main memory, or in temporary memory stores called data registers. At the same time, the control unit prepares the calculating unit for performing the actual calculations, such as comparing two pieces of information. When this has been done, the result may be stored, ready for the next operation. Or it may be sent to the output unit.

33 What language do computers use?

A computer can work on very complicated information. But despite this it can, in fact, only do two basic things. It can add two numbers together or compare two numbers to see if they are the same or not. Multiplication is only a series of additions, and subtraction and division are achieved by a clever manipulation of the numbers which are then added together.

A computer, therefore, uses a very simple language. Its switches and circuits can only be turned on or off, using either an electrical pulse or no pulse. This language is represented by just two digits, 1s (on) and 0s (off), known as binary code. A series of 1s and 0s can represent decimal numbers, as shown opposite. Each digit has twice the value of the one on its right, so the binary number 11010 means: $(1\times16)+(1\times8)+(0\times4)+(1\times2)+(0\times1)$, which is 26. Binary numbers can also represent letters.

Binary code can be recorded on magnetic tape or special disks, or as 'holes' or 'no holes' in punched cards or paper tape.

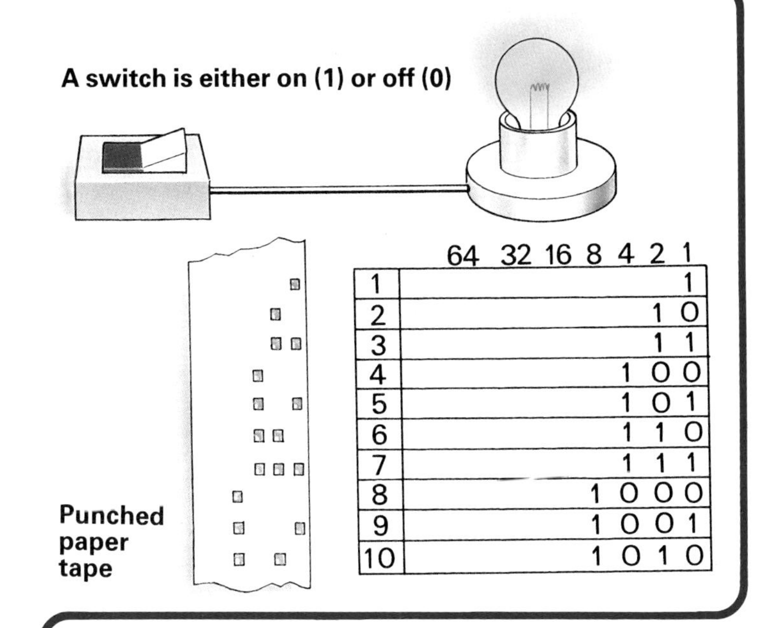

	64	32	16	8	4	2	1
1							1
2						1	0
3						1	1
4					1	0	0
5					1	0	1
6					1	1	0
7					1	1	1
8				1	0	0	0
9				1	0	0	1
10				1	0	1	0

34 What are bits and bytes?

A single 1 or 0 is known as a binary digit, or bit. The central processor of a computer handles information in groups of bits. These may consist of 4, 8, 16 or 32 bits, but most microcomputers handle 8-bit groups known as bytes. A byte may be sent along a single wire as a stream of pulses and spaces. Or, as shown here, each bit may be sent along eight parallel wires.

A kilobyte (K) is 1024 bytes. The size of a computer's memory is often described by the number of kilobytes it can hold, such as 16K, 32K or 64K.

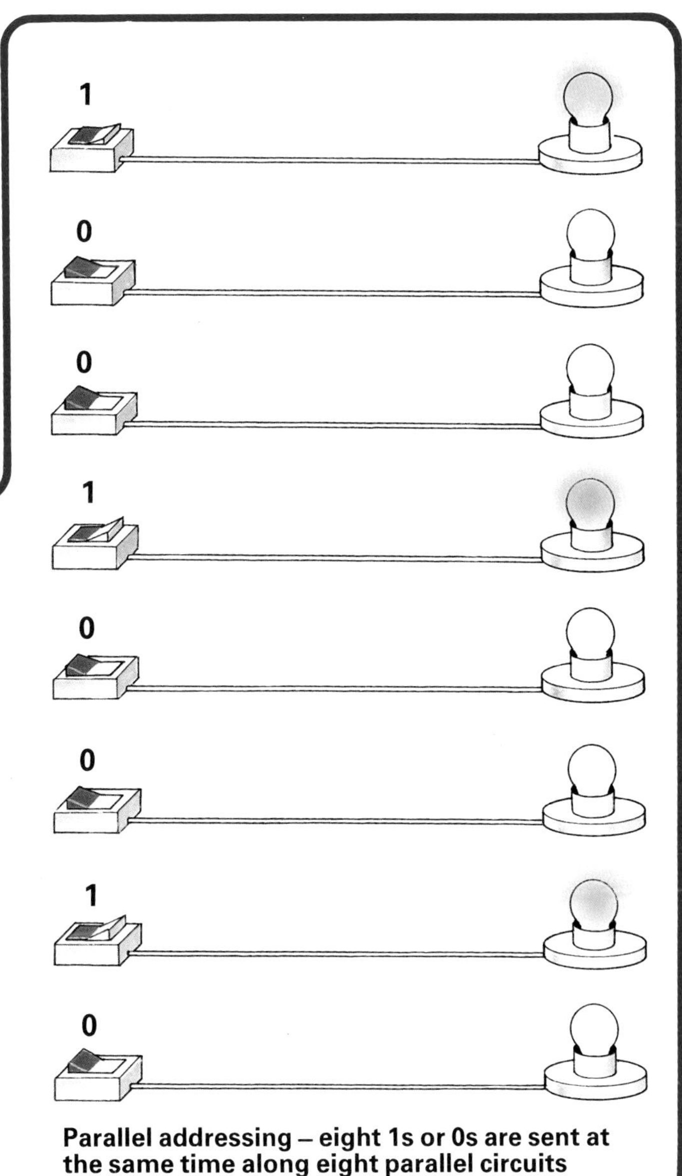

Parallel addressing – eight 1s or 0s are sent at the same time along eight parallel circuits

35 How does a computer remember?

Most computers have two kinds of memory – external and internal. External memory is for storing programs, or additional information, outside the computer. External memory comes in several forms. Early computers used punched cards or paper tape, but today, magnetic tapes and disks are much more common. Microcomputers often use tape cassettes or thin plastic disks called floppy disks. Larger computers use large reels of tape or hard disks.

Internal memories are 'printed' on silicon chips, of which there are two main kinds. Read only memories (ROMs) store information permanently. It cannot be changed or added to. ROMs are used to hold information that the computer needs all the time, such as how to add binary numbers.

Random access memories (RAMs), on the other hand, are used to store temporary information which is fed in from an external memory store, such as a computer game program you feed in from a cassette tape. A RAM consists of hundreds of tiny storage cells arranged in rows. Each one can receive a pulse of electricity and store it as an electrical charge. Types of storage cell vary. One type consists of a capacitor and a transistor. Another type is a special transistor that can store an electrical charge by itself.

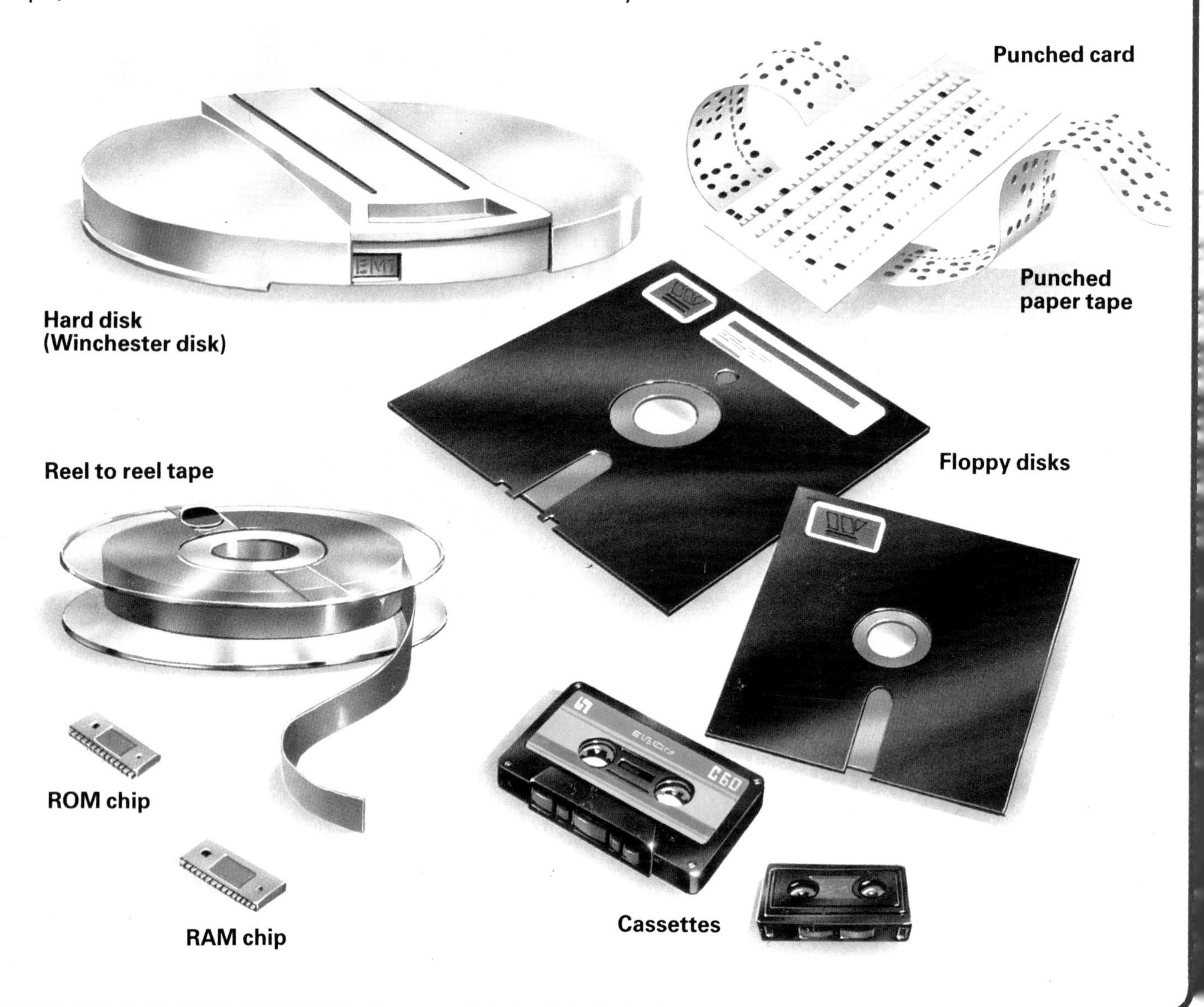

36 What does a databank do?

A databank is simply a computer with a very large memory store. Most of the memory is outside the computer in the form of hard disks or big reels of tapes.

Databanks are used by organizations that need to store a lot of information. Some libraries store catalogues and indexes in this way. Details of car registration numbers and drivers' licences are stored in databanks. And the police and other government organizations store information about people and events.

37 How do we talk to a computer?

The binary code used by computers is called machine language. But although this code is easy for a computer to handle, it is very difficult for people to use and mistakes are easily made. So computer experts have made up other languages for people to write programs in.

The simplest language is assembly language, in which binary numbers are represented by short codes of letters and decimal numbers. These are then translated by the computer into machine code. More advanced languages, known as high-level languages, come closer to our own language. Instructions to the computer can be typed in using a keyboard, and each letter and number on the keyboard has its own byte (8-bit) code. A translation program built into the computer's memory translates each instruction into machine code. The most common language used in microcomputers is called BASIC. Other computer languages include LOGO, which is used in schools and COBOL, which is used for business programs.

38 Are computers intelligent?

Computers are not intelligent. A computer has to be told exactly what to do by a very carefully worked out set of instructions, called a program. It cannot work things out for itself, or make 'spur of the moment' decisions in the way that we can. In fact, the 'intelligence' of a big modern computer is about the same as that of a dragonfly.

Even so, scientists are putting much time and effort into trying to develop intelligent computers. One of the main problems is that the human brain can hold a vast amount of information in its cells, far more than the biggest computer. In time, however, it may be possible to use a network of special chips that can hold this much information.

39 Can computers speak?

There are now a number of computers that make noises that can be recognized as speech. The Bell Telephone Company in the USA uses a speaking computer instead of a human directory-enquiries operator. There are also speaking chess sets and electronic teaching aids (see page 24).

In most speaking computers a 'dictionary' of word sounds is stored in the computer memory. It includes all the combinations of vowels and consonants that the computer needs. The sounds are fed into the computer by a person speaking into a microphone. The sound waves are analysed and converted into binary code, which is stored in the memory. To make a word the computer is programmed to select the right sounds from its memory and convert them into electrical signals, which are then fed to a loudspeaker.

40 Can computers read and listen?

A computer can be taught to analyse, store and recognize the patterns of words spoken by a particular person. But no two people speak in exactly the same way. So the computer will not recognize words spoken by anyone else.

At present, the best way around this problem is to 'train' a computer, using the voices of about 100 people. The computer stores an 'average' sound pattern for each word and will recognize the same words 80 to 90 times out of 100.

Computers can read printed words, using devices called optical character readers (OCRs). Each letter is scanned by a series of light sensors linked to a microprocessor. The shape of the letters is recorded as pulses (1s) or no pulses (0s) by the microprocessor and this information is compared with information stored in the computer memory.

Handwritten words are much more difficult for a computer to recognize. The most advanced readers can recognize handwritten capital letters. But it will be a long time before computers can cope with the enormous variation between people's joined-up handwriting.

Electronics Electronics

Electronics

Printed letter seen by array of light sensors

Information translated into binary code

```
0 0 0 0 0 0 0 0 0 0 0 0 0 0 0 0 0 0
0 1 1 1 1 1 0 1 1 1 1 1 1 1 0 0 0 0
0 0 1 1 1 1 1 1 0 0 0 1 1 1 1 0 0 0
0 0 0 1 1 1 1 0 0 0 0 0 1 1 1 0 0 0
0 0 0 1 1 1 0 0 0 0 0 0 1 1 1 0 0 0
0 0 0 1 1 1 0 0 0 0 0 0 1 1 1 0 0 0
0 0 0 1 1 1 0 0 0 0 0 0 1 1 1 0 0 0
0 0 0 1 1 1 0 0 0 0 0 0 1 1 1 0 0 0
0 0 0 1 1 1 0 0 0 0 0 0 1 1 1 0 0 0
0 0 0 1 1 1 0 0 0 0 0 0 1 1 1 0 0 0
0 0 0 1 1 1 0 0 0 0 0 0 1 1 1 0 0 0
0 0 0 1 1 1 0 0 0 0 0 0 1 1 1 0 0 0
0 0 0 1 1 1 0 0 0 0 0 0 1 1 1 0 0 0
0 0 0 1 1 1 0 0 0 0 0 0 1 1 1 0 0 0
0 0 0 1 1 1 0 0 0 0 0 0 1 1 1 0 0 0
0 0 1 1 1 1 1 0 0 0 0 1 1 1 1 1 0 0
0 1 1 1 1 1 1 1 0 0 1 1 1 1 1 1 1 0
0 0 0 0 0 0 0 0 0 0 0 0 0 0 0 0 0 0
```

41 Can computers draw?

Computers draw in the same way as they write. Binary information from the central processor instructs an electron gun to light up particular areas of the screen.

For the purpose of drawing, a computer screen is divided into areas known as pixels (picture cells). In some systems there may be only 800 pixels on the screen, resulting in 'low resolution' graphics in which pictures appear as blocks of small squares. The best 'high resolution' graphics use over 58,000 pixels.

A computer can be made to draw shapes by typing in the right instructions on a

keyboard. Programs for computer games are made in this way. It is also possible to draw directly onto a screen using a light pen. Another type of pen can be used on a special, sensitive graphics tablet. Nothing appears on the tablet, but lines and coloured areas appear on the screen as the pen moves. A system called FLAIR enables artists to create series of very detailed computer pictures. These can be put together to make films, such as cartoons. Computer pictures can also be recorded on paper, using a special plotter.

42 What is a word processor?

A word processor is a computer that is specially programmed to handle text. The text is typed in using a keyboard and appears on the screen. The program in the computer allows the operator to correct mistakes, to alter the spacing between words or lines and set margins on the screen. When the text is all correct it can be printed out or stored on a disk or tape.

This system is useful for producing large numbers of identical letters. If the computer is given the names and addresses of the people who are to receive the letters, it can automatically print out each letter with the correct name and address on it. Some word processors have additional programs that enable them to pick out possible spelling errors by comparing the typed words with words stored in memory. Word processors can, therefore, save a lot of time.

43 Can you learn from a computer?

Computers are being used more and more in schools. One reason for this is that they are becoming much more a part of our everyday lives. Pupils are being taught how to use and program computers so that they will be able to cope with the computerized offices and factories of the future.

Another equally important use of computers in schools is in teaching all kinds of other subjects. There are programs that can help in the teaching of writing, spelling, arithmetic, science and technical drawing.

At one time, the few schools that had computers used a number of terminals (a keyboard joined to a screen), which were all linked to a large mainframe computer. Today, many schools use separate microcomputers. These can be linked together like terminals, or used for individual teaching programs.

There are also a number of simple electronic teaching aids for young children. A spelling aid, for example, can store up to 240 words. The aid speaks out a word and the child spells it by pressing the right keys.

Electronic teaching aids

What are electronic games?

There is a wide range of small, portable electronic games, such as chess, battleship and backgammon. Like the teaching aids, they all use microprocessors to receive instructions from input units (usually a set of pushbuttons) and send instructions to output units like screens and loudspeakers.

But the most versatile games are those that can be fed into a computer from a tape or disk. Many of these can be played using microcomputers and television sets.

What is television information?

As well as normal programmes, many television sets can now receive specially-prepared information, such as news items and details about booking holidays or buying goods. There are two systems which supply this information, 'Teletext' and 'Viewdata'. In both cases, the information is prepared on a large central computer and then sent directly to your television screen.

Teletext information is prepared at the television company and broadcast with the usual television signal. To receive Teletext your television must have a special 'decoder' to decode the information signal. Viewdata is prepared by a group of business organizations and is sent to your screen through the telephone. For Viewdata you need a special device linking your telephone to the television. With both systems, you choose which item you want to see by pressing buttons on a small control console.

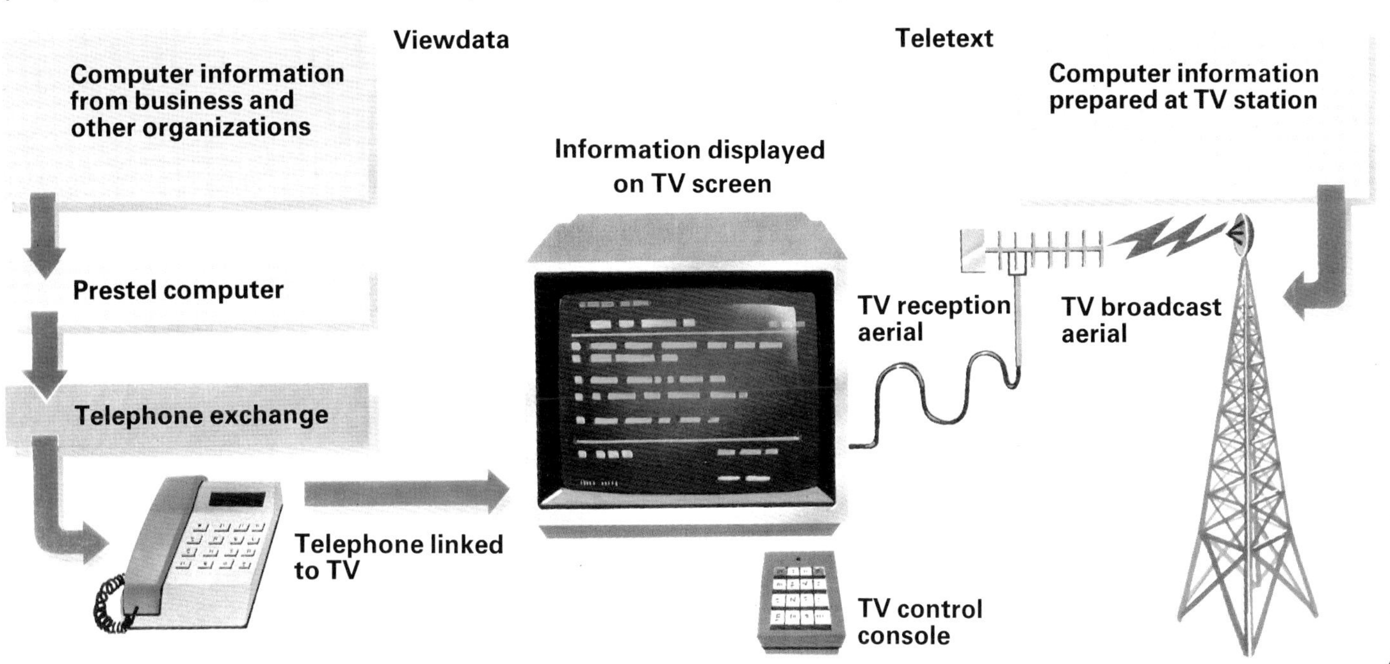

46 What is a robot?

In films and books, robots are electronic machines that act in similar ways to humans. Many have human-like bodies. But in the real world robots are not so far advanced. In fact the machines we call robots are simply computers with mechanical output devices – often mechanical arms. Most modern robots are industrial machines that carry out such tasks as welding, loading and paint spraying. And many robots can be programmed to do all these tasks.

Much work is being done to improve robots. Japanese scientists are making robots for use in fishing, forestry and medicine. Even a robot office secretary is planned and a robot sheep shearer has already been built. A few robots even have senses, such as sight, hearing and touch.

47 How do electronic telephones work?

As might be expected, electronics is taking over the telephone system. There are now several electronic telephone exchanges in the world and more and more electronic telephones are being used.

In an ordinary electrical telephone system, speech is carried along the wire as an audio wave – a varying electric current. In an electronic system the speech pattern is changed into electrical pulses. Each part of the audio wave is given a value in binary code. One advantage of this is that telephone lines can be used more easily for communicating with computers.

Electronic telephones have memories that can hold up to ten numbers and can automatically dial these numbers for you. In the future, videophones will allow callers to see as well as talk to each other.

A videophone conference

48 How is electronics used in aircraft?

Computers have been used in aircraft navigation systems and automatic pilots for some years. But the latest aircraft, such as the Boeing 757, have computers that monitor all of the aircraft systems, including the performance of the engines and their use of fuel. And instead of hundreds of instruments these aircraft have just a few video screens. Any information that the pilot wants can be displayed on a screen simply by pressing a button.

49 How is electronics used in cars?

Cars equipped with computers and electronic devices are already on the road. Solid-state (no moving parts) instrument panels show speed, rpm, mileage, fuel level, oil pressure and engine temperature on special illuminated displays. With the addition of a microprocessor the displays can also show such things as fuel consumption, average speed and estimated time of arrival. In the future, information from a device similar to a television tube may be projected onto the windscreen in what is called a head-up display (a system that is already used in some aircraft). As a result the driver will seldom have to take his eyes off the road ahead.

Computerizing a car involves three stages. First, sensors have to be placed at all the necessary points in the car. Second, the information from the sensors has to be turned into electrical pulses and fed to a microprocessor. Finally, instructions from the microprocessor have to be sent to the devices that control the working parts of the car. Electronic systems that control the ignition timing and fuel supply are already in use. In the future, braking, suspension and gearing systems will probably also be under computer control.

50 How is electronics used in space?

Without computers, modern space missions would be impossible. Using computers, scientists on the ground can place satellites in orbit and send spacecraft vast distances into space with pin-point accuracy.

All modern satellites and spacecraft carry their own on-board computers. These monitor and control the functions of all the working parts, such as rocket engines, radio equipment and cameras.

Computers and electronic devices will continue to play a vital part in space technology. In the future, robot spacecraft may be used to explore deep into space and to build space stations and cities (which will, of course, be under the control of computers when in use). One project that has been suggested is a lunar city where robots could mine metals from the Moon and process them in automated factories.

SWITCH – A Frustration Game!

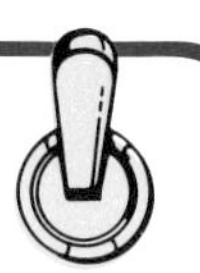

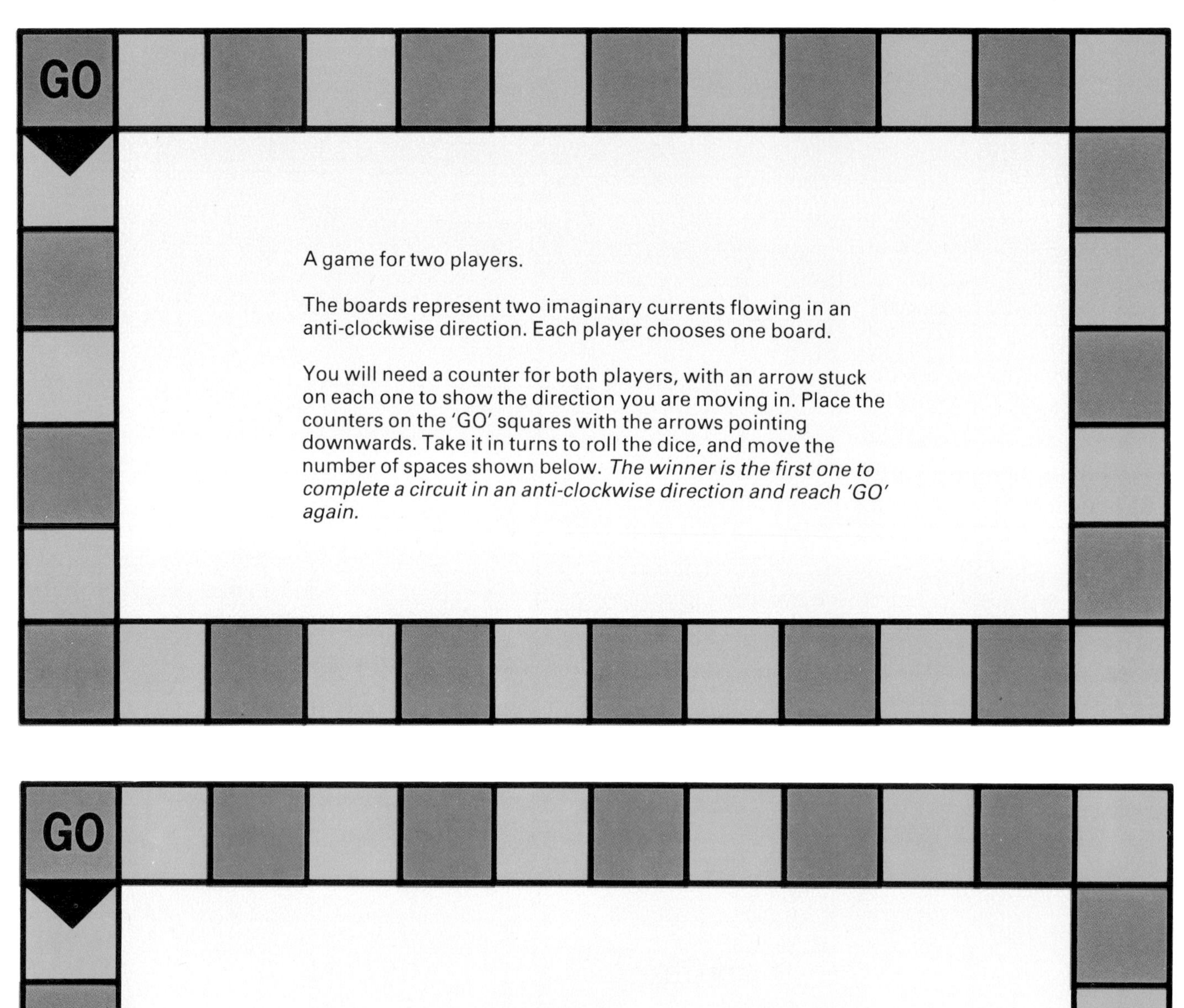

A game for two players.

The boards represent two imaginary currents flowing in an anti-clockwise direction. Each player chooses one board.

You will need a counter for both players, with an arrow stuck on each one to show the direction you are moving in. Place the counters on the 'GO' squares with the arrows pointing downwards. Take it in turns to roll the dice, and move the number of spaces shown below. *The winner is the first one to complete a circuit in an anti-clockwise direction and reach 'GO' again.*

GO

When a player throws a 1, 2, 3 or 4, they move the number of spaces shown on the dice. If they throw a 5 or a 6 they move 4 places only and shout 'SWITCH'. The other player must then turn their counter to face the opposite direction (clockwise). The other player must then move in that direction until their opponent throws a 5 or a 6 again and shouts 'SWITCH'. *Players only respond to their opponent's shout of 'SWITCH' – not their own.*

The Electronics Name Game

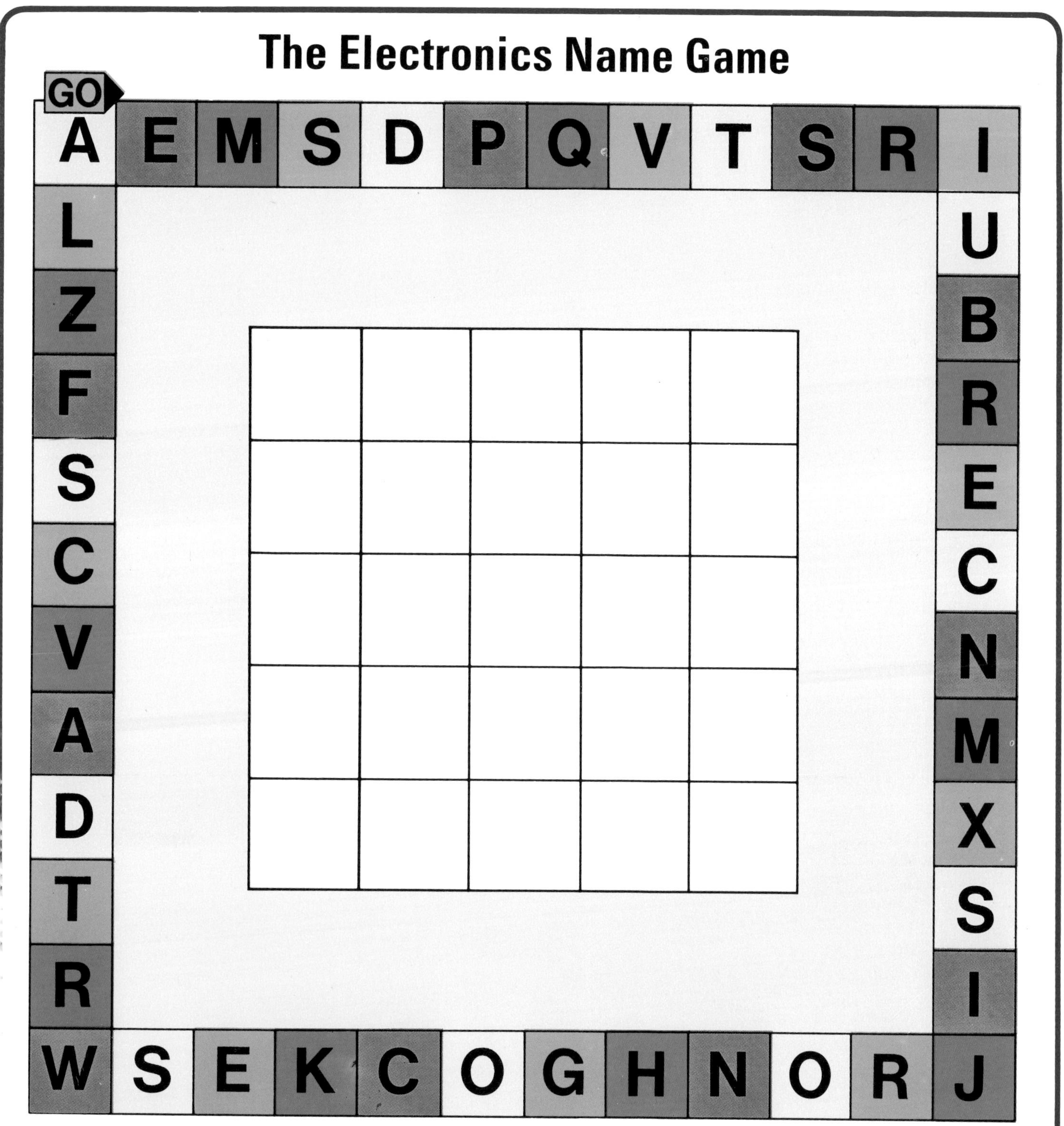

The Electronics Name Game

You will need a dice, coloured counters, pencils and paper. The players take turns in throwing the dice and moving their counters. Starting at 'GO', throw the dice and move your counter along the outer circle of squares by the number thrown. Then write down the letter of the square you have landed on. When you have enough letters to form the name of any person, electronic term, component or device mentioned in this book, write the name out and cross the letters used off your list. You can then put a counter on one of the squares in the centre box.

The aim is to complete a line of 5 counters (vertical, horizontal, or diagonal) in the centre box. You may block your opponent's line with one of your counters, but you may then find it more difficult to complete a line yourself. A player may use a name only once; the other player(s) may also use that name – but only once. The game may also be played so that the winner is the first to cover a block of six squares. Or simply carry on the game until all the squares in the centre box have been filled – then the player with the most counters is the winner.